SUDOKU

100 PUZZLES WITH SOLUTIONS

MULTI LEVEL **BOOK 10**

ISBN: 9781694063052

A standard Sudoku puzzle consists of a grid of 9 blocks. Each block contains 9 boxes arranged in 3 rows and 3 columns.

Column			Block			Box		
8	4	1	7	9	6		3	2
			2	5	4	8	1	
6	2	5	3	1		9	4	7
2		8		3	7	4	5	1
3		7	4	8		2	6	9
	5	6	9		1			8
5							2	3
1	6	3	8	4	2	7	9	5
9	7	2	5	6	3	1	8	4

(Row label at right of the fifth row)

The Basic Rules of Sudoku:

- There's only one solution to a Sudoku puzzle. A puzzle is considered solved when all 81 boxes contain numbers by following the Sudoku rules.
- When you start a game of Sudoku some blocks already have numbers (fewer number make a harder puzzle). These numbers cannot be changed.
- Each column must contain every number from 1 to 9 and no two numbers in the same column can be duplicated.
- Each row must contain every number from 1 to 9 and no two numbers in the same row can be duplicated.
- Each block must contain every number from 1 to 9 and no two numbers in the same block can be duplicated.

Here's the solved puzzle:

8	4	1	7	9	6	5	3	2
7	3	9	2	5	4	8	1	6
6	2	5	3	1	8	9	4	7
2	9	8	6	3	7	4	5	1
3	1	7	4	8	5	2	6	9
4	5	6	9	2	1	3	7	8
5	8	4	1	7	9	6	2	3
1	6	3	8	4	2	7	9	5
9	7	2	5	6	3	1	8	4

Puzzle 1
Easy

1		4	7	6			3	
9	5	6	8	3	2	7	1	4
7		3		4	1	6		
6	7		2	1				3
	4	2	6	5	3	9	7	1
3	1			7		2		6
2	6	1		8		3	9	7
4	9	8	3	2	7	1	6	5
5	3	7	1	9	6	4	2	8

Puzzle 2
Easy

		6				7	2	8
8	7	9	6	1		5		4
2	5	4	3	7				1
4	2	7	8	6	3	1	5	
6	9	5			1		8	
3		1	7	5			6	2
9	1	3	5		7	8	4	6
7	4	8	9	3	6	2	1	5
5	6	2	1	8	4	9	7	3

Puzzle 3
Easy

7	3	9	8	6	2	1	4	5
	1	2	9	4	5	3		7
	4		3	7	1	2		
5	7	4		2	9	8		
3	9	1	7	8	4	5	2	6
2	6	8	5		3	9	7	4
4	5	7	1	3	8		9	2
	8	6	2		7	4		
1	2			9	6	7	5	8

Puzzle 4
Easy

	3		5		4		9	8
4	6	8	1	3	9	2	7	5
7	5	9	6	8	2	1	4	
3	8	7		4	6			2
6				2	1	9	8	7
9	1	2		5	7	3	6	4
1	7	3	2	6	8	4	5	9
8		6	4	9		7	3	1
5	9					8	2	6

Puzzle 5
Easy

4	7	9	1	8			5	3
3	1	2	9	5		7		
8	6	5	7	3		9	1	
1	9	6	4	2	5	3	7	8
5	8	7	3	1	9			
2	4		6	7	8	5	9	1
7	2	4	5	6	1	8	3	9
9	5	8	2	4	3	1	6	7
6				9				

Puzzle 6
Easy

4			7	6	9	2	5	1
5	6	7	1	2	4	9	3	8
2	9	1	5	3	8			7
7	1		2	9			8	
6			8	1			7	9
8	5	9	4	7	6	3	1	2
3	7	6	9	4	1	8	2	5
1	4		6		2	7	9	3
9			3		7	1		

Puzzle 7
Easy

3	2	4		5		6		1
1	9	6		2	3	4	5	
8	7	5	4	6	1	2	9	3
5	6	8	1		7	3	2	
9	3	1			5	7		
7	4	2	6	3			1	5
2	1	7	3		6	5	4	
6	8	9	5	7	4	1	3	2
4	5	3		1	2			

Puzzle 8
Easy

9	4	2	7	8	5	3	6	1
	3	8	4	6	9	5	7	2
7	5	6			2			8
8	7	1				2		5
2	6	5			7		8	
3	9	4	5	2	8	6	1	
6	1	3	2	7	4	8	5	9
5	2	7	8	9		1		
4	8	9		5	1	7	2	

Puzzle 9
Easy

5	2		1	6		9	8	7
9	1	4	7	2	8	5	6	3
6	7	8	5	3	9	2	4	1
8	9	5	3	1	7	6		4
2	3		4	8	6		9	5
4	6		2		5	3	1	8
		9	6	4	1	8	3	2
3	4	6	8	5	2	1		
			9					6

Puzzle 10
Easy

1	3	5	9	7	2	6	4	8
4	9	8	1	6	5	3	2	7
7	6	2	3	4	8	5	9	1
9	8	7	4	5	1	2	3	6
6	4	1	2	9	3			5
5	2	3	6	8	7	4	1	9
	5			3		1		
	1			2	6			
	7			1				

Puzzle 11
Easy

7			1	6	4	2	9	5
1	2	9	7	5	8	6	4	3
			3	2	9	8		
8	9	7	4	1	2	5	3	6
2	5	6	8	9	3	4		
4			6	7	5	9	2	8
9	7	2	5	8	1	3	6	4
			9	4		7	5	2
			2	3	7	1	8	9

Puzzle 12
Easy

	4	8	6			3	9	5
	7	5	4		8	2	1	
2			3	1	5	7	8	4
6	8	7		3	1	4	5	9
5		1	7	4	9	8	6	2
4		2	5	8	6	1	3	
7	5	3	8	6		9		1
8	1	6				5		
9	2	4	1	5	3			8

Puzzle 13
Easy

7	4	1	2	3	6		9	8
8	2	6	9	5	7	1	3	4
	3		1	4	8	7	6	2
6	7	4	8	9		3	2	1
2	1	8	6	7	3	9	4	5
9	5	3	4		1	6	8	
3		7	5		4	2	1	
1		5	3		2			
4	6	2						

Puzzle 14
Easy

3		4	8	5	7			
8	1	5	9		4	2	7	3
		9	3	1	2	8		4
9	5	8		2		6	4	7
7	4	1	6	8	5	3	2	9
2	6	3	7	4	9	5	1	8
	9	6	5	7		1		
	3	2	4	9	1			
	8	7	2	3	6	4	9	5

Puzzle 15
Easy

2	5	3	9		8	6		4
8	7		5	3	6		2	1
6	9	1	4	7	2			5
5	2	7	1	6	4			
3	1	8		5	9	4	6	2
9		6	2	8	3	1	5	
4	3	5	8	9		2	1	6
7	8	9	6	2		5	4	
1	6	2	3	4	5	7		

Puzzle 16
Easy

2		1	6	5	9	4	8	
5	4	8		7		1	9	6
6	7	9	1	8	4	3	2	5
			5	6		8	7	4
4	8			9		5	6	1
	5			4		9	3	2
	1	4	9	2		6	5	3
	6			1	5	2	4	9
9	2	5	4	3	6	7	1	8

Puzzle 17
Easy

5	9	4	6	8				
8	6	1	7	2	4	9	5	3
7	2		5		9	8	6	4
9	5	7	3	4				6
4	8	2	9	6		3		5
1	3	6	2	5		4	9	
3	1		4	7	6	5		
	7	8	1	3		6	4	
6	4		8	9	2	7	3	

Puzzle 18
Easy

2	8	5	1	4		7	6	3
4		7	8				9	2
1	9	3		6	2	8	4	5
	7	1		8	5		2	4
	5	6	4	2			7	8
		4		9				1
6	3	2	9	1	4	5	8	7
7	4	8	5	3		2	1	9
5	1	9	2	7	8	4	3	6

Puzzle 19
Easy

4	3	1	6	2		5	9	7
2	7	6	5	9	3	1	8	4
	5	8		4	7		2	3
1	9	5		6	2	4		8
6	2	4		8	1			9
3		7		5	4	2	6	1
7	1	9	2	3	5	8	4	6
8	6	2	4		9			5
5	4	3	8		6	9		

Puzzle 20
Easy

7	8	4	9	1	3	2	6	5
2	3	5	7	8	6	4	1	9
9	6	1	4	2	5	7	8	3
	2	7		6	9		4	8
	4	8		7	2			6
	9	6	1	4	8		2	7
8					4		7	1
6	1	2	8		7			4
4			6	5	1			2

Puzzle 21
Easy

1	5	2	6	9	3	7	8	4
7		8				3	9	
4	3	9	7	8			6	1
8	2	1	5	7	9	4	3	6
6	9	5		3		1	7	2
3	7	4			6	9	5	8
5	8	3	9	4		6		7
2		6	3	5	7	8		9
9		7		6				

Puzzle 22
Easy

9	8		4		2	3	5	1
3	6	5		9	8			2
2		1	7			6		9
7	5	4	2	8	1	9	3	6
1	3	9		7			2	8
		8	9			1	7	4
8	1	6	3	4	7	2	9	5
5	9	3	8	2	6	4	1	
4	7	2	5		9	8	6	3

Puzzle 23
Easy

9	5		7	8		4	1	3
8	1			2		9	6	
3	6			9	1		2	
7		3	9		5	6	4	
6	4			7	2	5	3	9
	9			6			7	8
5	2	6	1	3	8	7	9	4
4	3	9	6	5	7	2	8	1
1	7	8	2	4	9	3	5	6

Puzzle 24
Easy

5	1	2	3	6		8		4
9	8	3		1			2	
6	7	4	9	8				3
	2	9	4		1	6	7	5
3		1	2	7	5			8
4	5		8			3	1	2
1	3	5		4	8	2	6	9
2	9		1	5	3	4		7
	4	8	6	2	9	5	3	1

Puzzle 25
Easy

	3	8	6	1	4	7	9	5
7	5	1	9	3	2	4	8	6
6	9	4	5	8	7	2	1	3
4	7	9	1	5		8		2
	2	6	4	7	8		3	1
8		3	2			5		
		2		4	1		5	8
1		5	3	2	9	6		
3	4	7	8	6	5	1	2	9

Puzzle 26
Medium

4	2						3	
6	1			3	8	4	2	
3			2				1	
9					3	2	4	1
2		4				9	7	3
1	7	3	4	2	9	6	5	8
7	4	6	3	8	2	1	9	5
8	9	1				3	6	2
5	3	2	9	6	1	7	8	4

Puzzle 27
Medium

8	4	3	1	2	5	7	6	
9	1	7	6	3	4	8	5	2
		5				4	1	
7	5	8	4	1	2	3	9	6
4	3	9	5			2	8	
				8		5	7	4
						9	3	
3	9	4	7	5	1	6	2	8
5	8						4	7

Puzzle 28
Medium

7	3	4	5	1	9			6
1	2	6	8	4	3	9	7	5
9	5	8	6	7	2	1	3	4
6	7	2	9	8	4	5	1	3
3	9	5	7	6	1			2
4	8	1	3	2	5			
2		3						8
		9						
		7						

Puzzle 29
Medium

6	3			4	7	1	8	9
1	2	9			3	5	4	7
8	7	4	5	9	1	3	6	2
3	8		9		4	7		
						9		
5				7				
	1	8				4	9	
		7	4	3	2	6	1	8
4	6	3	8	1	9	2	7	5

Puzzle 30
Medium

7	9			4	6			
5	4	1	2			6	3	7
8	6			7	1		4	
1	7	9	4	6	2	8	5	3
4	3	5			7	2		
2	8	6		3	5		7	4
9	1	7			3	4		
6	2	4	7			3		
3	5	8			4	7		

Puzzle 31
Medium

5	8	7					1	
4	3	9	2	8	1	6	7	5
6	2	1	9		7	4	3	8
7				1	9		8	
8				7				1
1			8			7		
2	6	5	1	4	3	8	9	7
9	7	8				1	4	3
3	1	4	7	9	8	5	2	6

Puzzle 32
Medium

6	1	2		5	8			
8	5	7	4	6	3	1	9	2
9	4	3	1		2	8	6	5
7	3	9	5	8	6	2	1	
1		4	2	3	9	5		
2		5	7	1				9
3	7	6	8	4	5	9	2	1
4			6				5	
5			3					

Puzzle 33
Medium

8			9		2		5	1
4		5		1	6	7	8	2
6	2	1	8	7	5			
	1	6	2	5	9	8	4	7
	4	2	6	8	7			
5	7	8	4	3	1			6
2	8			6	4			
	5	9		2	3	4		8
		4		9	8			

Puzzle 34
Medium

1		6		3			5	7
3		4			7	1	6	9
7		2	1				3	
4	3	5				7		8
8	2	1	7	5	3	4	9	
9	6	7	2		8	5	1	3
5		8	3	1	9	6	4	2
2	4	3		8		9	7	1
6	1	9	4					

Puzzle 35
Medium

	3	7				2	6	8
	6	8	3	7		1	4	9
	4	9				7	5	3
8		1				3	7	
7		3				6	9	
4	9	6	2	3	7	8	1	5
6	7	5					3	1
3	1	2	7				8	6
9	8	4			3	5	2	7

Puzzle 36
Medium

			4	6	2	8	9	7
7	2	8	1	9	5	4	3	6
		4	8	3	7	5	1	2
2					3			
3		5	7	2	1	9	6	
						2		3
			3	5	8	7	2	9
			2	7	4	6	8	1
8	7	2	6	1	9	3	4	5

Puzzle 37
Medium

	6	1	4	3	8		5	2
8	3	4	5		2		6	
5			1		6	3	8	4
2	4	3	6	5				8
6	9	8	7	4	1	5	2	3
1	5	7			3	6	4	9
3		5		6	4	8		
4		6			7	2		5
					5	4		6

Puzzle 38
Medium

	2	9	4	3	5	8	7	1
4	3	1				9	6	
8	5	7	9	1	6	3		2
			2		3		9	8
5	8	6	1	4		2	3	7
	9	3	8					4
1	6			8	4	7	5	
3							2	6
9	4	5		7	2			

Puzzle 39
Medium

2	5	6	7	9	1	4		
	8	4	3		6			
9		1			8	6		5
5		2	8	7	3			
6	1	7	9	4	5	8	2	3
	9		1	6	2	5	4	7
	7		5	1	4			
1	2		6				5	4
4	6	5	2	3	9	7	8	1

Puzzle 40
Medium

	8		6			3	9	
			8	9			6	
	6	9	7			8		
		6	1	3	8	2	4	9
9	4	8	2	7	6			
1	3	2	5	4	9	6	7	8
	1		4	6	7	9	8	2
8	9	7	3	2	1	4	5	
6	2	4	9	8	5			

Puzzle 41
Medium

4			8	7	9		2	5
		6		1	3	8		7
	8	7	5	6	2	1	3	
7	6	9		8	4	3		1
	1	8	9	5	6	4	7	2
			7	3	1		6	
		3		2	5	7		9
	9		3	4	7			
	7	2	1	9	8	5	4	3

Puzzle 42
Medium

2	5	8	9	4	1			
	4	3	7	2	6	8	5	1
6	1	7				2	9	4
5	9	4	6	1	2			8
3	8	6		5	7	9	1	2
1	7		8			6		5
		1	2			5		9
		5	1		9	4		
		9	5			1		

Puzzle 43
Medium

3	6	4	2	1	7	8		
		8	3	9			6	
	2	9	5	6	8	4	3	
2			9	8	6	5	4	3
	9		1	4	3	7	2	8
4	8	3	7	5	2			
	4		8	7	9			
9				3				
8	3		6	2		9	7	4

Puzzle 44
Medium

7	4	9				8	6	2
	6		7	2	9	4	5	1
5	1	2	6	4		3	7	9
9	3	5	4	1	6	7		8
4		1				6	3	5
6	2	7	8	5	3	1	9	4
	7					2	4	
2	5		3			9	1	
	9					5	8	

Puzzle 45
Medium

9			6			2	1	8
8	1	5	7	3		6	9	4
2	6		8	1	9	3	7	5
4	8		5	7	6	9	3	1
1	9	6	3			5		7
3	5		9	4	1	8		
5		9		8		1	2	
	4		2		5	7	8	3
7	2				3			

Puzzle 46
Medium

8	4	5	6			2		3
7		6		8	3	5		4
	3	1			2	7	8	6
	8				6		4	
	6				1	3	5	8
1	5	7	8	3	4	6	2	9
5	9	2	3		8	4		1
			2	1	9	8		5
	1	8			5	9		2

Puzzle 47
Medium

						7	6	
3	6				7	5	8	
7	2	9	6	5	8	4	1	3
4		2				6	9	
			9	2	6	3		4
	9	3			5	1	2	
5	3	6	8	7	9	2	4	1
9			3	6	2	8		
2	8	7	5	1	4	9	3	6

Puzzle 48
Medium

1		2				3	7	
		7				4	5	
4		5				1	6	
	1	6	7	9	5	4	8	2
5	4	9	2	6		1	7	3
2	7	8				5	6	9
7	5	3	4	8	9	6	2	1
8	2	1				3	9	4
9	6	4				7	5	8

Puzzle 49
Medium

6	4	7		3	9		5	
				5	7		6	4
5	1		4	8	6		7	
7	5	6		1	2	4	8	9
		8	6		5	7	1	2
1	2	4	7	9	8	6	3	5
	6		5	7			9	1
		5			1		2	7
	7	1		2	3	5	4	6

Puzzle 50
Medium

	6			1		8		7
	1		9		8		4	5
4	8	9	3	5	7	6		2
5	7	6	1	3	9	2	8	4
9	4					1	7	6
1	2	8	7	4	6	5	3	9
6	9					7	5	1
8	5		6	7		9		3
	3			9		4		8

Puzzle 51
Hard

	3	8	2	7	9	5	6	
			4					3
	1	9		3			4	7
	4	3	8	9				6
	5			2		4		8
	8			4				5
		4				3	8	9
3		5	9	8		6	1	4
8	9	1	3	6	4	7	5	

Puzzle 52
Hard

	5	8	6	3			2	1
	1		5					3
	9		2	1				
9	3		4	6	2	1		8
8	4	1	3			2	5	6
		6	8	5	1			
5	8	9	7	4	3		1	2
		4	9	2	5			
		7	1	8			9	

Puzzle 53
Hard

				2				3
2				7				8
3	4	6		1		7	2	5
5			4	8		2		6
	2			6	1	5	4	7
6	1	4	7	5	2	3	8	9
4			2	9			3	1
1		2		3				4
	3		1	4			5	2

Puzzle 54
Hard

4		7	3					
		1		7				
		3	9	8				
		9		1				
		4		6		1	9	
7	1	8	2	9	3	5	4	6
8	4	5	3	2	9	7	6	1
1	7	2	6	5	8	4	3	9
9	3	6		4				

Puzzle 55
Hard

			5	3	1	2	7	8
	5	3	7	4		6	9	1
	7			6	9	3	5	4
5			3	8	6	9	4	7
6	3	7	9	5	4	8	1	2
	8		1	2	7	5	3	6
		5	6		3			9

Puzzle 56
Hard

4	6	7				9	1	
2	1	9						4
8	3	5	1	9	4	2	6	7
1						8		6
9	8					1		
3		6	8	1		4	9	
5	4	1						9
	2	8		3			4	
	9	3	5	4				

Puzzle 57
Hard

8		6	1	5	4	7	2	
		5						
1	7	9	3					4
7	9	8	4			6		
3	5	2	9	6				
6				7	3	2		
	2		6	4	1	5		7
	6	1					4	2
4		7	2	3	5	9	1	6

Puzzle 58
Hard

	5	6				8	9	2
			2	8		6		
	8	2	9	6		3	1	
		7			6	1		
	4		1		3		2	6
	3		7	4	2	5	8	9
			4	1		9		
3		9		2	8	4		1
	1	4		3	9	2		

Puzzle 59
Hard

6	8	2	4	3	1	7	5	9
7	5	1	2	9	8	3		
9	4	3	5	7	6	1	8	2
8	6	5	7	4	9	2	1	3
		9	8			6		5
					5			
1		6				5		
3				5				
5								

Puzzle 60
Hard

8		3			6	7		
	2	9	3	5	7	4	6	8
7		6	2		8	3	5	
	8				3		7	6
9	7	2				8		5
6	3	1	7	8	5	2		4
	6			7	2		1	3
2		8				6		
3							8	2

Puzzle 61
Hard

9	3		7	5				
							9	
			9					
6	7	3	1	4	5	8	2	
5	9	2	8	3	6			
	4	1	2	9	7			
3			4	2	1	9		7
7	2	9	3			5		
1	6	4	5	7	9			

Puzzle 62
Hard

6		2		8	4	1	7	5
				5	1	4		
5	4	1	6	7		3		8
4		5		9				
	2		1	6	3			4
		6	7	4	5	8		2
			5	2			4	
2		4		1				
	5		4	3	6	2	8	1

Puzzle 63
Hard

5	7	1		4	6	8		2
8	4		3	1	2	9	7	5
2	9	3				4		6
9	5	8	7	2				3
6		7	1		8	2		
1	2	4	6	3	9	5	8	7
						6		
	6		4		1			
	1							

Puzzle 64
Hard

7			2					
2	9		7		6			
6				8				2
5	7				1	8		6
3	8	6		7				1
4	1	2		8	9	5	3	7
1	6	3	4	5	2	7	8	9
8			9		7	1		3
9		7	8		3			4

Puzzle 65
Hard

7	6	8	2	5	9		4	
3	2	5		1			9	6
9	4	1	6	3				
6	7	9	1	4	3			5
4	5				2	6	1	7
8	1	2	7	6	5	9	3	4
	9	6	3	2				

Puzzle 66
Hard

		5	2				6	9
6					5			
2			6				5	
	3		9					
5	6	7	4	1	2	9	8	3
9	4	2	5	8	3	7	1	6
		1	3			5		
3	5	6				8	9	
4	2	9	8	5				

Puzzle 67
Hard

6	7	8	1		3	4	5	
			8	6	7			
			4					7
7	3	6	9	8	4	5	2	1
4	5	2	3	1	6	9	7	8
8	1	9	5		2	3	4	6
	6				8		9	
	8			5				

Puzzle 68
Hard

	8	1	5	7	3	6		
		5	9	2	4	8	1	7
			6	8	1	5	3	
		8		1	7			6
			4	9	8		5	1
	1			6	5			
8	9	3	7	4	2	1	6	5
1			8					
			1				8	

Puzzle 69
Hard

	2		9	1	7	6	3	8
9	3	7		6	2	4	1	5
1	6	8			5	9	2	7
			5	8	3		9	
	5	9			6	8	7	
	8	2			9		5	
					8		4	
					4		8	
8	4	3			1	5	6	9

Puzzle 70
Hard

1			6	9	3			
6			5	2	1			
3	5		8	7	4			
5			3	4	2	1		
			7	1	5			
4	3	1		8	6			
7	6	3	2	5	9			
	1		4	6	7	3		5
	4	5	1	3	8	7	6	9

Puzzle 71
Hard

6	3					5	9	2
	9		6				3	8
	2						6	1
2	5	9				8	1	
8	6	7	5	1	9	3	2	4
1	4	3				9	5	
	8	2		6	7	1	4	
3	7	6	1			2	8	
	1					6	7	

Puzzle 72
Hard

4			3	7				1
		3	1			4		
	1	7	8		4		2	3
2	5	1	4	3	6			
3	4		2	8	7	1	5	
7		8	9	1	5	2	3	4
9					3	8	1	
1		6	7		8		4	
					1			

Puzzle 73
Hard

7			5		9	6	1	3
6	9	3	4	8	1			
	2	1		3	7			
			9		6	1		
1				4	3			
			7	1	8			
3	1	6	8		4		9	
	5		3	6	2			1
2			1	9	5	3	6	8

Puzzle 74
Hard

7	2	9		1		6	5	8
6			7	8	9	4	2	1
8	1	4				3	7	9
9		6		7				
	8					7	6	
		7						3
5	7	8			1	2	3	6
4		2		3				7
3		1			7	5	8	4

Puzzle 75
Hard

	1	3	4		7		8	
	6	4			8		7	
8	2	7			5		1	
		1		7		8		
		5			2	1		7
7		2						6
4	3	9			6	7		1
1	7	8					6	9
2	5	6	7	9	1	3	4	8

Puzzle 76
Fiendish

3	9	1				5	6	8
2	8	6	5	9	1			
7	5	4	6		8	2	1	9
			3	1	5			4
4			8	6	9			
	1		7	2	4	8		
			9					

Puzzle 77
Fiendish

1		8			5	9	3	4
5	9	3	4	1	8	2	6	7
4			3	9				
				6		3		9
	3					4		
		1						
3			1				9	
							4	6
		7			9			3

Puzzle 78
Fiendish

						3		5
	4			3	8	2		
2	3	6	5	7	9	4	1	8
					7			
	5		3				2	1
	6					1		2
	2	8		1		6	5	
9	1	7				8		

Puzzle 79
Fiendish

2			4	1	9	6	3	7
		3				5	2	9
9	6	7	3	2	5			
	7							5
3						2		1
						3		8
6					7			2
7								6
		1	6			7	4	3

Puzzle 80
Fiendish

		3	8	7	1	9	6	
			3	5	2	8	4	
8	7	5			4	2	1	3
						6		9
	8							4
				2		3		1
		7					9	6
			2			1		8
		8			5			2

Puzzle 81
Fiendish

8		6	4	1			2	9
			6	7		4	1	3
4			9		2			8
6		9				8	4	1
7		8				9	3	6
						2	5	7
5				2				4
								2
						1	9	5

Puzzle 82
Fiendish

1	4						3	
					3		8	
3	8				5		1	
						5	2	3
					6	4		8
	3					1		6
4		8					6	
						3	4	
6	7	3	9	2	4	8	5	1

Puzzle 83
Fiendish

		3		9				
1	9			3				
			1	6		9	3	
	5		3	2	9			7
	3		4		1			
9			6		8			
			2	4	3	5	9	1
		9		1				
	1	5	9	8				

Puzzle 84
Fiendish

8	1					4		9
9	6							
3	2				9	7		
5	9	6	7	8	1			
2	4	3	9	6			7	
7	8	1		2	4			5
	5			9	8		3	
	3		2		7			
	7							

Puzzle 85
Fiendish

	8	7	1	6	2	5		
	6	5	9	3	4			
	4		7	8	5			
8	5	6	2	9	3	4	1	7
						6	2	9
			6			3	5	8
	9							
	2	8						5
					9			

Puzzle 86
Fiendish

4	6					3	5	9
3	9							
1	8			3	9		6	4
6		9	4		3	5	2	8
2					8	4	1	6
			2	6		9	3	7
7	2	6	3	9				

Puzzle 87
Fiendish

2		8	6	5	7			
5		9					7	
4		7				5		
6	9	5						7
	7	2	5				4	
	8	4		7			2	5
9		6			1	7		3
7	4	3	9					
8		1	7					

Puzzle 88
Fiendish

	8		7				5	
	5				4	2		7
	4			3	5			
8		5	4		3			
3								
1	9	2	5	6	8	7		
	3					8	2	1
	1				2	9		
	2	8						6

Puzzle 89
Fiendish

3	1	2	8		9		6	4
	4	9	1					
			4					
				4				
2	5	6	3		8	4		
		4	5			6		
4			2					
	2		9		4			
9	6	3	7	8	1	2	4	5

Puzzle 90
Fiendish

6	8	5	7				1	3
			9	3		8	4	
	9	3			8	2	7	
		8					3	
3				7		5	9	
5	1	7				6	2	4
							8	
	3	1				4		
8							5	

Puzzle 91
Fiendish

			4			2	8	3
	7		9			1	4	6
					2	7	5	9
				8		5	3	
1			5	4		9	7	
7						4	6	
							2	4
					4		9	7
	4					8	1	5

Puzzle 92
Fiendish

8	1		5			3	9	2
	7		2					
	6	2					7	
2	3	6				7	5	4
4	9	8	7	2		6	1	3
	5	7	3			2	8	9
							2	
			6		2	5		
	2							

Puzzle 93
Fiendish

		8					7	9
5		2		7	9		3	
		7	1	6		5		
		4	3		7			
		6	5	2				
2		3	6		4	7		
3	4		7					
8	2		9					7
6	7							

Puzzle 94
Fiendish

4	3		2	8	6	5		
	2	8	7	5	1	3	9	4
7	1	5	4	9	3			
			6		9			5
	9	7	5		4		3	
	4	6	8	3	7			
	5		9					

Puzzle 95
Fiendish

	4	7						
5	8		2					
	9				7			
	1							
	2			6				
9	6	5						
4	5	9	6	2	8	1		
	7		3	5	4			9
8	3	2	9	7	1	4	6	5

Puzzle 96
Fiendish

3	4	6	8	7	5	2	1	9
5	2	7	3			4	6	8
9	8	1	6			5		
7	5						2	6
						7		
						9	8	
								2
6				8		3		1
								7

Puzzle 97
Fiendish

			1	5				9
	9		4	7	2			
	5		9	8				
	6		5	1	7		3	
			2	6	9		4	
			8	3	4	9		
			6	2		7		3
	7		3	9			1	4
	3		7	4				

Puzzle 98
Fiendish

				9		1		
			1	2				
3	9	1		5		2	6	
4	7		6		9	5		
					5		4	
			9	3	2	4	5	6
			5	4	8	7		2
2	5	4	7	6	1	3		

Puzzle 99
Fiendish

4	3	5	8	7	1			
8	9	2	6		5	7		
6	1	7	3	2	9			8
	8			9	3	1		
1	5			6	4			
	2			1	8			
				3				
5		1	4	8				

Puzzle 100
Fiendish

6	3	9			1	4	5	
5						1		6
1					5			
						5		
					8			
2			7	3	6			
4		3		1	2			
7					3			4
8		6	9	4	7	3	1	5

Solutions

Puzzle 1

1	2	4	7	6	5	8	3	9
9	5	6	8	3	2	7	1	4
7	8	3	9	4	1	6	5	2
6	7	9	2	1	8	5	4	3
8	4	2	6	5	3	9	7	1
3	1	5	4	7	9	2	8	6
2	6	1	5	8	4	3	9	7
4	9	8	3	2	7	1	6	5
5	3	7	1	9	6	4	2	8

Puzzle 2

1	3	6	4	9	5	7	2	8
8	7	9	6	1	2	5	3	4
2	5	4	3	7	8	6	9	1
4	2	7	8	6	3	1	5	9
6	9	5	2	4	1	3	8	7
3	8	1	7	5	9	4	6	2
9	1	3	5	2	7	8	4	6
7	4	8	9	3	6	2	1	5
5	6	2	1	8	4	9	7	3

Puzzle 3

7	3	9	8	6	2	1	4	5
6	1	2	9	4	5	3	8	7
8	4	5	3	7	1	2	6	9
5	7	4	6	2	9	8	1	3
3	9	1	7	8	4	5	2	6
2	6	8	5	1	3	9	7	4
4	5	7	1	3	8	6	9	2
9	8	6	2	5	7	4	3	1
1	2	3	4	9	6	7	5	8

Puzzle 4

2	3	1	5	7	4	6	9	8
4	6	8	1	3	9	2	7	5
7	5	9	6	8	2	1	4	3
3	8	7	9	4	6	5	1	2
6	4	5	3	2	1	9	8	7
9	1	2	8	5	7	3	6	4
1	7	3	2	6	8	4	5	9
8	2	6	4	9	5	7	3	1
5	9	4	7	1	3	8	2	6

Puzzle 5

4	7	9	1	8	2	6	5	3
3	1	2	9	5	6	7	8	4
8	6	5	7	3	4	9	1	2
1	9	6	4	2	5	3	7	8
5	8	7	3	1	9	2	4	6
2	4	3	6	7	8	5	9	1
7	2	4	5	6	1	8	3	9
9	5	8	2	4	3	1	6	7
6	3	1	8	9	7	4	2	5

Puzzle 6

4	3	8	7	6	9	2	5	1
5	6	7	1	2	4	9	3	8
2	9	1	5	3	8	4	6	7
7	1	3	2	9	5	6	8	4
6	2	4	8	1	3	5	7	9
8	5	9	4	7	6	3	1	2
3	7	6	9	4	1	8	2	5
1	4	5	6	8	2	7	9	3
9	8	2	3	5	7	1	4	6

Puzzle 7

3	2	4	7	5	9	6	8	1
1	9	6	8	2	3	4	5	7
8	7	5	4	6	1	2	9	3
5	6	8	1	9	7	3	2	4
9	3	1	2	4	5	7	6	8
7	4	2	6	3	8	9	1	5
2	1	7	3	8	6	5	4	9
6	8	9	5	7	4	1	3	2
4	5	3	9	1	2	8	7	6

Puzzle 8

9	4	2	7	8	5	3	6	1
1	3	8	4	6	9	5	7	2
7	5	6	1	3	2	4	9	8
8	7	1	9	4	6	2	3	5
2	6	5	3	1	7	9	8	4
3	9	4	5	2	8	6	1	7
6	1	3	2	7	4	8	5	9
5	2	7	8	9	3	1	4	6
4	8	9	6	5	1	7	2	3

Puzzle 9

5	2	3	1	6	4	9	8	7
9	1	4	7	2	8	5	6	3
6	7	8	5	3	9	2	4	1
8	9	5	3	1	7	6	2	4
2	3	1	4	8	6	7	9	5
4	6	7	2	9	5	3	1	8
7	5	9	6	4	1	8	3	2
3	4	6	8	5	2	1	7	9
1	8	2	9	7	3	4	5	6

Puzzle 10

1	3	5	9	7	2	6	4	8
4	9	8	1	6	5	3	2	7
7	6	2	3	4	8	5	9	1
9	8	7	4	5	1	2	3	6
6	4	1	2	9	3	7	8	5
5	2	3	6	8	7	4	1	9
2	5	6	8	3	9	1	7	4
8	1	4	7	2	6	9	5	3
3	7	9	5	1	4	8	6	2

Puzzle 11

7	3	8	1	6	4	2	9	5
1	2	9	7	5	8	6	4	3
5	6	4	3	2	9	8	1	7
8	9	7	4	1	2	5	3	6
2	5	6	8	9	3	4	7	1
4	1	3	6	7	5	9	2	8
9	7	2	5	8	1	3	6	4
3	8	1	9	4	6	7	5	2
6	4	5	2	3	7	1	8	9

Puzzle 12

1	4	8	6	2	7	3	9	5
3	7	5	4	9	8	2	1	6
2	6	9	3	1	5	7	8	4
6	8	7	2	3	1	4	5	9
5	3	1	7	4	9	8	6	2
4	9	2	5	8	6	1	3	7
7	5	3	8	6	2	9	4	1
8	1	6	9	7	4	5	2	3
9	2	4	1	5	3	6	7	8

Puzzle 13

7	4	1	2	3	6	5	9	8
8	2	6	9	5	7	1	3	4
5	3	9	1	4	8	7	6	2
6	7	4	8	9	5	3	2	1
2	1	8	6	7	3	9	4	5
9	5	3	4	2	1	6	8	7
3	8	7	5	6	4	2	1	9
1	9	5	3	8	2	4	7	6
4	6	2	7	1	9	8	5	3

Puzzle 14

3	2	4	8	5	7	9	6	1
8	1	5	9	6	4	2	7	3
6	7	9	3	1	2	8	5	4
9	5	8	1	2	3	6	4	7
7	4	1	6	8	5	3	2	9
2	6	3	7	4	9	5	1	8
4	9	6	5	7	8	1	3	2
5	3	2	4	9	1	7	8	6
1	8	7	2	3	6	4	9	5

Puzzle 15

2	5	3	9	1	8	6	7	4
8	7	4	5	3	6	9	2	1
6	9	1	4	7	2	3	8	5
5	2	7	1	6	4	8	3	9
3	1	8	7	5	9	4	6	2
9	4	6	2	8	3	1	5	7
4	3	5	8	9	7	2	1	6
7	8	9	6	2	1	5	4	3
1	6	2	3	4	5	7	9	8

Puzzle 16

2	3	1	6	5	9	4	8	7
5	4	8	2	7	3	1	9	6
6	7	9	1	8	4	3	2	5
3	9	2	5	6	1	8	7	4
4	8	7	3	9	2	5	6	1
1	5	6	7	4	8	9	3	2
8	1	4	9	2	7	6	5	3
7	6	3	8	1	5	2	4	9
9	2	5	4	3	6	7	1	8

Puzzle 17

5	9	4	6	8	3	1	2	7
8	6	1	7	2	4	9	5	3
7	2	3	5	1	9	8	6	4
9	5	7	3	4	8	2	1	6
4	8	2	9	6	1	3	7	5
1	3	6	2	5	7	4	9	8
3	1	9	4	7	6	5	8	2
2	7	8	1	3	5	6	4	9
6	4	5	8	9	2	7	3	1

Puzzle 18

2	8	5	1	4	9	7	6	3
4	6	7	8	5	3	1	9	2
1	9	3	7	6	2	8	4	5
3	7	1	6	8	5	9	2	4
9	5	6	4	2	1	3	7	8
8	2	4	3	9	7	6	5	1
6	3	2	9	1	4	5	8	7
7	4	8	5	3	6	2	1	9
5	1	9	2	7	8	4	3	6

Puzzle 19

4	3	1	6	2	8	5	9	7
2	7	6	5	9	3	1	8	4
9	5	8	1	4	7	6	2	3
1	9	5	3	6	2	4	7	8
6	2	4	7	8	1	3	5	9
3	8	7	9	5	4	2	6	1
7	1	9	2	3	5	8	4	6
8	6	2	4	1	9	7	3	5
5	4	3	8	7	6	9	1	2

Puzzle 20

7	8	4	9	1	3	2	6	5
2	3	5	7	8	6	4	1	9
9	6	1	4	2	5	7	8	3
1	2	7	3	6	9	5	4	8
3	4	8	5	7	2	1	9	6
5	9	6	1	4	8	3	2	7
8	5	3	2	9	4	6	7	1
6	1	2	8	3	7	9	5	4
4	7	9	6	5	1	8	3	2

Puzzle 21

1	5	2	6	9	3	7	8	4
7	6	8	2	1	4	3	9	5
4	3	9	7	8	5	2	6	1
8	2	1	5	7	9	4	3	6
6	9	5	4	3	8	1	7	2
3	7	4	1	2	6	9	5	8
5	8	3	9	4	1	6	2	7
2	1	6	3	5	7	8	4	9
9	4	7	8	6	2	5	1	3

Puzzle 22

9	8	7	4	6	2	3	5	1
3	6	5	1	9	8	7	4	2
2	4	1	7	3	5	6	8	9
7	5	4	2	8	1	9	3	6
1	3	9	6	7	4	5	2	8
6	2	8	9	5	3	1	7	4
8	1	6	3	4	7	2	9	5
5	9	3	8	2	6	4	1	7
4	7	2	5	1	9	8	6	3

Puzzle 23

9	5	2	7	8	6	4	1	3
8	1	4	5	2	3	9	6	7
3	6	7	4	9	1	8	2	5
7	8	3	9	1	5	6	4	2
6	4	1	8	7	2	5	3	9
2	9	5	3	6	4	1	7	8
5	2	6	1	3	8	7	9	4
4	3	9	6	5	7	2	8	1
1	7	8	2	4	9	3	5	6

Puzzle 24

5	1	2	3	6	7	8	9	4
9	8	3	5	1	4	7	2	6
6	7	4	9	8	2	1	5	3
8	2	9	4	3	1	6	7	5
3	6	1	2	7	5	9	4	8
4	5	7	8	9	6	3	1	2
1	3	5	7	4	8	2	6	9
2	9	6	1	5	3	4	8	7
7	4	8	6	2	9	5	3	1

Puzzle 25

2	3	8	6	1	4	7	9	5
7	5	1	9	3	2	4	8	6
6	9	4	5	8	7	2	1	3
4	7	9	1	5	3	8	6	2
5	2	6	4	7	8	9	3	1
8	1	3	2	9	6	5	4	7
9	6	2	7	4	1	3	5	8
1	8	5	3	2	9	6	7	4
3	4	7	8	6	5	1	2	9

Puzzle 26

4	2	8	1	9	6	5	3	7
6	1	7	5	3	8	4	2	9
3	5	9	2	4	7	8	1	6
9	6	5	8	7	3	2	4	1
2	8	4	6	1	5	9	7	3
1	7	3	4	2	9	6	5	8
7	4	6	3	8	2	1	9	5
8	9	1	7	5	4	3	6	2
5	3	2	9	6	1	7	8	4

Puzzle 27

8	4	3	1	2	5	7	6	9
9	1	7	6	3	4	8	5	2
2	6	5	8	7	9	4	1	3
7	5	8	4	1	2	3	9	6
4	3	9	5	6	7	2	8	1
1	2	6	9	8	3	5	7	4
6	7	1	2	4	8	9	3	5
3	9	4	7	5	1	6	2	8
5	8	2	3	9	6	1	4	7

Puzzle 28

7	3	4	5	1	9	2	8	6
1	2	6	8	4	3	9	7	5
9	5	8	6	7	2	1	3	4
6	7	2	9	8	4	5	1	3
3	9	5	7	6	1	8	4	2
4	8	1	3	2	5	6	9	7
2	4	3	1	9	6	7	5	8
8	6	9	4	5	7	3	2	1
5	1	7	2	3	8	4	6	9

Puzzle 29

6	3	5	2	4	7	1	8	9
1	2	9	6	8	3	5	4	7
8	7	4	5	9	1	3	6	2
3	8	1	9	2	4	7	5	6
7	4	6	3	5	8	9	2	1
5	9	2	1	7	6	8	3	4
2	1	8	7	6	5	4	9	3
9	5	7	4	3	2	6	1	8
4	6	3	8	1	9	2	7	5

Puzzle 30

7	9	2	3	4	6	5	1	8
5	4	1	2	8	9	6	3	7
8	6	3	5	7	1	9	4	2
1	7	9	4	6	2	8	5	3
4	3	5	8	1	7	2	6	9
2	8	6	9	3	5	1	7	4
9	1	7	6	2	3	4	8	5
6	2	4	7	5	8	3	9	1
3	5	8	1	9	4	7	2	6

Puzzle 31

5	8	7	4	3	6	2	1	9
4	3	9	2	8	1	6	7	5
6	2	1	9	5	7	4	3	8
7	4	6	5	1	9	3	8	2
8	5	2	3	7	4	9	6	1
1	9	3	8	6	2	7	5	4
2	6	5	1	4	3	8	9	7
9	7	8	6	2	5	1	4	3
3	1	4	7	9	8	5	2	6

Puzzle 32

6	1	2	9	5	8	4	7	3
8	5	7	4	6	3	1	9	2
9	4	3	1	7	2	8	6	5
7	3	9	5	8	6	2	1	4
1	6	4	2	3	9	5	8	7
2	8	5	7	1	4	6	3	9
3	7	6	8	4	5	9	2	1
4	2	1	6	9	7	3	5	8
5	9	8	3	2	1	7	4	6

Puzzle 33

8	3	7	9	4	2	6	5	1
4	9	5	3	1	6	7	8	2
6	2	1	8	7	5	3	9	4
3	1	6	2	5	9	8	4	7
9	4	2	6	8	7	1	3	5
5	7	8	4	3	1	9	2	6
2	8	3	1	6	4	5	7	9
1	5	9	7	2	3	4	6	8
7	6	4	5	9	8	2	1	3

Puzzle 34

1	8	6	9	3	4	2	5	7
3	5	4	8	2	7	1	6	9
7	9	2	1	6	5	8	3	4
4	3	5	6	9	1	7	2	8
8	2	1	7	5	3	4	9	6
9	6	7	2	4	8	5	1	3
5	7	8	3	1	9	6	4	2
2	4	3	5	8	6	9	7	1
6	1	9	4	7	2	3	8	5

Puzzle 35

1	3	7	5	9	4	2	6	8
5	6	8	3	7	2	1	4	9
2	4	9	1	6	8	7	5	3
8	2	1	9	5	6	3	7	4
7	5	3	4	8	1	6	9	2
4	9	6	2	3	7	8	1	5
6	7	5	8	2	9	4	3	1
3	1	2	7	4	5	9	8	6
9	8	4	6	1	3	5	2	7

Puzzle 36

1	5	3	4	6	2	8	9	7
7	2	8	1	9	5	4	3	6
6	9	4	8	3	7	5	1	2
2	6	7	9	4	3	1	5	8
3	8	5	7	2	1	9	6	4
9	4	1	5	8	6	2	7	3
4	1	6	3	5	8	7	2	9
5	3	9	2	7	4	6	8	1
8	7	2	6	1	9	3	4	5

Puzzle 37

7	6	1	4	3	8	9	5	2
8	3	4	5	9	2	1	6	7
5	2	9	1	7	6	3	8	4
2	4	3	6	5	9	7	1	8
6	9	8	7	4	1	5	2	3
1	5	7	8	2	3	6	4	9
3	7	5	2	6	4	8	9	1
4	1	6	9	8	7	2	3	5
9	8	2	3	1	5	4	7	6

Puzzle 38

6	2	9	4	3	5	8	7	1
4	3	1	7	2	8	9	6	5
8	5	7	9	1	6	3	4	2
7	1	4	2	5	3	6	9	8
5	8	6	1	4	9	2	3	7
2	9	3	8	6	7	5	1	4
1	6	2	3	8	4	7	5	9
3	7	8	5	9	1	4	2	6
9	4	5	6	7	2	1	8	3

Puzzle 39

2	5	6	7	9	1	4	3	8
7	8	4	3	5	6	2	1	9
9	3	1	4	2	8	6	7	5
5	4	2	8	7	3	1	9	6
6	1	7	9	4	5	8	2	3
3	9	8	1	6	2	5	4	7
8	7	3	5	1	4	9	6	2
1	2	9	6	8	7	3	5	4
4	6	5	2	3	9	7	8	1

Puzzle 40

4	8	1	6	5	2	3	9	7
2	7	3	8	9	4	1	6	5
5	6	9	7	1	3	8	2	4
7	5	6	1	3	8	2	4	9
9	4	8	2	7	6	5	1	3
1	3	2	5	4	9	6	7	8
3	1	5	4	6	7	9	8	2
8	9	7	3	2	1	4	5	6
6	2	4	9	8	5	7	3	1

Puzzle 41

4	3	1	8	7	9	6	2	5
2	5	6	4	1	3	8	9	7
9	8	7	5	6	2	1	3	4
7	6	9	2	8	4	3	5	1
3	1	8	9	5	6	4	7	2
5	2	4	7	3	1	9	6	8
1	4	3	6	2	5	7	8	9
8	9	5	3	4	7	2	1	6
6	7	2	1	9	8	5	4	3

Puzzle 42

2	5	8	9	4	1	3	6	7
9	4	3	7	2	6	8	5	1
6	1	7	3	8	5	2	9	4
5	9	4	6	1	2	7	3	8
3	8	6	4	5	7	9	1	2
1	7	2	8	9	3	6	4	5
7	3	1	2	6	4	5	8	9
8	2	5	1	3	9	4	7	6
4	6	9	5	7	8	1	2	3

Puzzle 43

3	6	4	2	1	7	8	5	9
7	5	8	3	9	4	1	6	2
1	2	9	5	6	8	4	3	7
2	1	7	9	8	6	5	4	3
5	9	6	1	4	3	7	2	8
4	8	3	7	5	2	6	9	1
6	4	2	8	7	9	3	1	5
9	7	1	4	3	5	2	8	6
8	3	5	6	2	1	9	7	4

Puzzle 44

7	4	9	1	3	5	8	6	2
3	6	8	7	2	9	4	5	1
5	1	2	6	4	8	3	7	9
9	3	5	4	1	6	7	2	8
4	8	1	9	7	2	6	3	5
6	2	7	8	5	3	1	9	4
8	7	3	5	9	1	2	4	6
2	5	6	3	8	4	9	1	7
1	9	4	2	6	7	5	8	3

Puzzle 45

9	7	3	6	5	4	2	1	8
8	1	5	7	3	2	6	9	4
2	6	4	8	1	9	3	7	5
4	8	2	5	7	6	9	3	1
1	9	6	3	2	8	5	4	7
3	5	7	9	4	1	8	6	2
5	3	9	4	8	7	1	2	6
6	4	1	2	9	5	7	8	3
7	2	8	1	6	3	4	5	9

Puzzle 46

8	4	5	6	9	7	2	1	3
7	2	6	1	8	3	5	9	4
9	3	1	4	5	2	7	8	6
3	8	9	5	2	6	1	4	7
2	6	4	9	7	1	3	5	8
1	5	7	8	3	4	6	2	9
5	9	2	3	6	8	4	7	1
4	7	3	2	1	9	8	6	5
6	1	8	7	4	5	9	3	2

Puzzle 47

8	1	5	2	4	3	7	6	9
3	6	4	1	9	7	5	8	2
7	2	9	6	5	8	4	1	3
4	5	2	7	3	1	6	9	8
1	7	8	9	2	6	3	5	4
6	9	3	4	8	5	1	2	7
5	3	6	8	7	9	2	4	1
9	4	1	3	6	2	8	7	5
2	8	7	5	1	4	9	3	6

Puzzle 48

1	8	2	5	4	6	9	3	7
6	3	7	9	1	2	8	4	5
4	9	5	8	7	3	2	1	6
3	1	6	7	9	5	4	8	2
5	4	9	2	6	8	1	7	3
2	7	8	1	3	4	5	6	9
7	5	3	4	8	9	6	2	1
8	2	1	6	5	7	3	9	4
9	6	4	3	2	1	7	5	8

Puzzle 49

6	4	7	1	3	9	2	5	8
3	8	9	2	5	7	1	6	4
5	1	2	4	8	6	9	7	3
7	5	6	3	1	2	4	8	9
9	3	8	6	4	5	7	1	2
1	2	4	7	9	8	6	3	5
2	6	3	5	7	4	8	9	1
4	9	5	8	6	1	3	2	7
8	7	1	9	2	3	5	4	6

Puzzle 50

3	6	5	2	1	4	8	9	7
2	1	7	9	6	8	3	4	5
4	8	9	3	5	7	6	1	2
5	7	6	1	3	9	2	8	4
9	4	3	8	2	5	1	7	6
1	2	8	7	4	6	5	3	9
6	9	2	4	8	3	7	5	1
8	5	4	6	7	1	9	2	3
7	3	1	5	9	2	4	6	8

Puzzle 51

4	3	8	2	7	9	5	6	1
5	2	7	4	1	6	8	9	3
6	1	9	5	3	8	2	4	7
7	4	3	8	9	5	1	2	6
9	5	6	1	2	3	4	7	8
1	8	2	6	4	7	9	3	5
2	6	4	7	5	1	3	8	9
3	7	5	9	8	2	6	1	4
8	9	1	3	6	4	7	5	2

Puzzle 52

4	5	8	6	3	7	9	2	1
6	1	2	5	9	4	7	8	3
7	9	3	2	1	8	4	6	5
9	3	5	4	6	2	1	7	8
8	4	1	3	7	9	2	5	6
2	7	6	8	5	1	3	4	9
5	8	9	7	4	3	6	1	2
1	6	4	9	2	5	8	3	7
3	2	7	1	8	6	5	9	4

Puzzle 53

7	9	8	5	2	4	1	6	3
2	5	1	3	7	6	4	9	8
3	4	6	8	1	9	7	2	5
5	7	9	4	8	3	2	1	6
8	2	3	9	6	1	5	4	7
6	1	4	7	5	2	3	8	9
4	6	5	2	9	7	8	3	1
1	8	2	6	3	5	9	7	4
9	3	7	1	4	8	6	5	2

Puzzle 54

4	8	7	1	3	6	9	2	5
6	9	1	5	7	2	3	8	4
2	5	3	9	8	4	6	1	7
3	6	9	4	1	5	2	7	8
5	2	4	8	6	7	1	9	3
7	1	8	2	9	3	5	4	6
8	4	5	3	2	9	7	6	1
1	7	2	6	5	8	4	3	9
9	3	6	7	4	1	8	5	2

Puzzle 55

4	6	9	5	3	1	2	7	8
2	5	3	7	4	8	6	9	1
1	7	8	2	6	9	3	5	4
5	1	2	3	8	6	9	4	7
6	3	7	9	5	4	8	1	2
9	8	4	1	2	7	5	3	6
7	2	5	6	1	3	4	8	9
3	4	1	8	9	2	7	6	5
8	9	6	4	7	5	1	2	3

Puzzle 56

4	6	7	2	5	3	9	1	8
2	1	9	6	7	8	5	3	4
8	3	5	1	9	4	2	6	7
1	5	4	3	2	9	8	7	6
9	8	2	4	6	7	1	5	3
3	7	6	8	1	5	4	9	2
5	4	1	7	8	6	3	2	9
6	2	8	9	3	1	7	4	5
7	9	3	5	4	2	6	8	1

Puzzle 57

8	3	6	1	5	4	7	2	9
2	4	5	7	8	9	1	6	3
1	7	9	3	2	6	8	5	4
7	9	8	4	1	2	6	3	5
3	5	2	9	6	8	4	7	1
6	1	4	5	7	3	2	9	8
9	2	3	6	4	1	5	8	7
5	6	1	8	9	7	3	4	2
4	8	7	2	3	5	9	1	6

Puzzle 58

1	5	6	3	7	4	8	9	2
9	7	3	2	8	1	6	4	5
4	8	2	9	6	5	3	1	7
2	9	7	8	5	6	1	3	4
5	4	8	1	9	3	7	2	6
6	3	1	7	4	2	5	8	9
8	2	5	4	1	7	9	6	3
3	6	9	5	2	8	4	7	1
7	1	4	6	3	9	2	5	8

Puzzle 59

6	8	2	4	3	1	7	5	9
7	5	1	2	9	8	3	4	6
9	4	3	5	7	6	1	8	2
8	6	5	7	4	9	2	1	3
4	1	9	8	2	3	6	7	5
2	3	7	1	6	5	4	9	8
1	9	6	3	8	4	5	2	7
3	2	4	9	5	7	8	6	1
5	7	8	6	1	2	9	3	4

Puzzle 60

8	5	3	1	4	6	7	2	9
1	2	9	3	5	7	4	6	8
7	4	6	2	9	8	3	5	1
4	8	5	9	2	3	1	7	6
9	7	2	4	6	1	8	3	5
6	3	1	7	8	5	2	9	4
5	6	4	8	7	2	9	1	3
2	1	8	5	3	9	6	4	7
3	9	7	6	1	4	5	8	2

Puzzle 61

9	3	6	7	5	2	1	4	8
2	1	5	6	8	4	7	9	3
4	8	7	9	1	3	2	5	6
6	7	3	1	4	5	8	2	9
5	9	2	8	3	6	4	7	1
8	4	1	2	9	7	6	3	5
3	5	8	4	2	1	9	6	7
7	2	9	3	6	8	5	1	4
1	6	4	5	7	9	3	8	2

Puzzle 62

6	3	2	9	8	4	1	7	5
8	7	9	3	5	1	4	2	6
5	4	1	6	7	2	3	9	8
4	1	5	2	9	8	7	6	3
7	2	8	1	6	3	9	5	4
3	9	6	7	4	5	8	1	2
1	8	3	5	2	7	6	4	9
2	6	4	8	1	9	5	3	7
9	5	7	4	3	6	2	8	1

Puzzle 63

5	7	1	9	4	6	8	3	2
8	4	6	3	1	2	9	7	5
2	9	3	5	8	7	4	1	6
9	5	8	7	2	4	1	6	3
6	3	7	1	5	8	2	4	9
1	2	4	6	3	9	5	8	7
4	8	5	2	7	3	6	9	1
3	6	2	4	9	1	7	5	8
7	1	9	8	6	5	3	2	4

Puzzle 64

7	3	5	1	2	4	6	9	8
2	9	8	7	3	6	4	1	5
6	4	1	5	9	8	3	7	2
5	7	9	3	4	1	8	2	6
3	8	6	2	7	5	9	4	1
4	1	2	6	8	9	5	3	7
1	6	3	4	5	2	7	8	9
8	2	4	9	6	7	1	5	3
9	5	7	8	1	3	2	6	4

Puzzle 65

7	6	8	2	5	9	1	4	3
3	2	5	4	1	7	8	9	6
9	4	1	6	3	8	5	7	2
6	7	9	1	4	3	2	8	5
4	5	3	8	9	2	6	1	7
8	1	2	7	6	5	9	3	4
1	9	6	3	2	4	7	5	8
2	3	7	5	8	1	4	6	9
5	8	4	9	7	6	3	2	1

Puzzle 66

7	1	5	2	3	8	4	6	9
6	8	4	1	9	5	3	7	2
2	9	3	6	4	7	1	5	8
1	3	8	9	7	6	2	4	5
5	6	7	4	1	2	9	8	3
9	4	2	5	8	3	7	1	6
8	7	1	3	6	9	5	2	4
3	5	6	7	2	4	8	9	1
4	2	9	8	5	1	6	3	7

Puzzle 67

6	7	8	1	2	3	4	5	9
9	4	5	8	6	7	1	3	2
1	2	3	4	9	5	6	8	7
7	3	6	9	8	4	5	2	1
4	5	2	3	1	6	9	7	8
8	1	9	5	7	2	3	4	6
5	6	1	2	4	8	7	9	3
3	8	7	6	5	9	2	1	4
2	9	4	7	3	1	8	6	5

Puzzle 68

4	8	1	5	7	3	6	2	9
3	6	5	9	2	4	8	1	7
2	7	9	6	8	1	5	3	4
5	3	8	2	1	7	9	4	6
6	2	7	4	9	8	3	5	1
9	1	4	3	6	5	2	7	8
8	9	3	7	4	2	1	6	5
1	4	2	8	5	6	7	9	3
7	5	6	1	3	9	4	8	2

Puzzle 69

4	2	5	9	1	7	6	3	8
9	3	7	8	6	2	4	1	5
1	6	8	4	3	5	9	2	7
7	1	4	5	8	3	2	9	6
3	5	9	1	2	6	8	7	4
6	8	2	7	4	9	1	5	3
2	9	6	3	5	8	7	4	1
5	7	1	6	9	4	3	8	2
8	4	3	2	7	1	5	6	9

Puzzle 70

1	7	4	6	9	3	2	5	8
6	8	9	5	2	1	4	3	7
3	5	2	8	7	4	6	9	1
5	9	7	3	4	2	1	8	6
8	2	6	7	1	5	9	4	3
4	3	1	9	8	6	5	7	2
7	6	3	2	5	9	8	1	4
9	1	8	4	6	7	3	2	5
2	4	5	1	3	8	7	6	9

Puzzle 71

6	3	4	7	8	1	5	9	2
7	9	1	6	2	5	4	3	8
5	2	8	9	4	3	7	6	1
2	5	9	4	3	6	8	1	7
8	6	7	5	1	9	3	2	4
1	4	3	2	7	8	9	5	6
9	8	2	3	6	7	1	4	5
3	7	6	1	5	4	2	8	9
4	1	5	8	9	2	6	7	3

Puzzle 72

4	8	2	3	7	9	5	6	1
5	9	3	1	6	2	4	7	8
6	1	7	8	5	4	9	2	3
2	5	1	4	3	6	7	8	9
3	4	9	2	8	7	1	5	6
7	6	8	9	1	5	2	3	4
9	7	5	6	4	3	8	1	2
1	2	6	7	9	8	3	4	5
8	3	4	5	2	1	6	9	7

Puzzle 73

7	4	8	5	2	9	6	1	3
6	9	3	4	8	1	7	2	5
5	2	1	6	3	7	8	4	9
4	3	2	9	5	6	1	8	7
1	8	7	2	4	3	9	5	6
9	6	5	7	1	8	2	3	4
3	1	6	8	7	4	5	9	2
8	5	9	3	6	2	4	7	1
2	7	4	1	9	5	3	6	8

Puzzle 74

7	2	9	3	1	4	6	5	8
6	3	5	7	8	9	4	2	1
8	1	4	2	5	6	3	7	9
9	4	6	5	7	3	8	1	2
1	8	3	4	9	2	7	6	5
2	5	7	1	6	8	9	4	3
5	7	8	9	4	1	2	3	6
4	6	2	8	3	5	1	9	7
3	9	1	6	2	7	5	8	4

Puzzle 75

5	1	3	4	6	7	9	8	2
9	6	4	1	2	8	5	7	3
8	2	7	9	3	5	6	1	4
3	4	1	6	7	9	8	2	5
6	9	5	8	4	2	1	3	7
7	8	2	5	1	3	4	9	6
4	3	9	2	8	6	7	5	1
1	7	8	3	5	4	2	6	9
2	5	6	7	9	1	3	4	8

Puzzle 76

3	9	1	2	4	7	5	6	8
2	8	6	5	9	1	4	7	3
7	5	4	6	3	8	2	1	9
8	6	7	3	1	5	9	2	4
4	2	3	8	6	9	1	5	7
5	1	9	7	2	4	8	3	6
1	3	8	9	7	2	6	4	5
6	4	5	1	8	3	7	9	2
9	7	2	4	5	6	3	8	1

Puzzle 77

1	6	8	7	2	5	9	3	4
5	9	3	4	1	8	2	6	7
4	7	2	3	9	6	8	5	1
7	2	4	5	6	1	3	8	9
6	3	5	9	8	7	4	1	2
9	8	1	2	3	4	6	7	5
3	4	6	1	5	2	7	9	8
2	1	9	8	7	3	5	4	6
8	5	7	6	4	9	1	2	3

Puzzle 78

7	8	1	4	6	2	3	9	5
5	4	9	1	3	8	2	6	7
2	3	6	5	7	9	4	1	8
1	9	2	8	4	7	5	3	6
8	5	4	3	9	6	7	2	1
6	7	3	2	5	1	9	8	4
3	6	5	9	8	4	1	7	2
4	2	8	7	1	3	6	5	9
9	1	7	6	2	5	8	4	3

Puzzle 79

2	5	8	4	1	9	6	3	7
1	4	3	7	6	8	5	2	9
9	6	7	3	2	5	1	8	4
8	7	2	9	3	1	4	6	5
3	9	6	5	8	4	2	7	1
4	1	5	2	7	6	3	9	8
6	3	4	8	5	7	9	1	2
7	2	9	1	4	3	8	5	6
5	8	1	6	9	2	7	4	3

Puzzle 80

2	4	3	8	7	1	9	6	5
6	9	1	3	5	2	8	4	7
8	7	5	6	9	4	2	1	3
1	5	2	4	8	3	6	7	9
3	8	9	7	1	6	5	2	4
7	6	4	5	2	9	3	8	1
5	2	7	1	3	8	4	9	6
9	3	6	2	4	7	1	5	8
4	1	8	9	6	5	7	3	2

Puzzle 81

8	7	6	4	1	3	5	2	9
2	9	5	6	7	8	4	1	3
4	1	3	9	5	2	6	7	8
6	2	9	5	3	7	8	4	1
7	5	8	2	4	1	9	3	6
1	3	4	8	9	6	2	5	7
5	6	1	3	2	9	7	8	4
9	4	7	1	8	5	3	6	2
3	8	2	7	6	4	1	9	5

Puzzle 82

1	4	6	7	8	9	2	3	5
5	9	7	2	1	3	6	8	4
3	8	2	6	4	5	9	1	7
7	6	4	1	9	8	5	2	3
2	1	9	3	5	6	4	7	8
8	3	5	4	7	2	1	9	6
4	2	8	5	3	1	7	6	9
9	5	1	8	6	7	3	4	2
6	7	3	9	2	4	8	5	1

Puzzle 83

5	6	3	7	9	2	4	1	8
1	9	2	8	3	4	7	5	6
4	7	8	1	6	5	9	3	2
6	5	4	3	2	9	1	8	7
8	3	7	4	5	1	2	6	9
9	2	1	6	7	8	3	4	5
7	8	6	2	4	3	5	9	1
2	4	9	5	1	6	8	7	3
3	1	5	9	8	7	6	2	4

Puzzle 84

8	1	5	6	7	3	4	2	9
9	6	7	5	4	2	3	1	8
3	2	4	8	1	9	7	5	6
5	9	6	7	8	1	2	4	3
2	4	3	9	6	5	8	7	1
7	8	1	3	2	4	6	9	5
6	5	2	4	9	8	1	3	7
1	3	8	2	5	7	9	6	4
4	7	9	1	3	6	5	8	2

Puzzle 85

3	8	7	1	6	2	5	9	4
1	6	5	9	3	4	7	8	2
2	4	9	7	8	5	1	3	6
8	5	6	2	9	3	4	1	7
4	3	1	5	7	8	6	2	9
9	7	2	6	4	1	3	5	8
5	9	3	4	2	6	8	7	1
6	2	8	3	1	7	9	4	5
7	1	4	8	5	9	2	6	3

Puzzle 86

4	6	2	7	8	1	3	5	9
3	9	5	6	4	2	8	7	1
1	8	7	5	3	9	2	6	4
6	7	9	4	1	3	5	2	8
2	5	3	9	7	8	4	1	6
8	1	4	2	6	5	9	3	7
7	2	6	3	9	4	1	8	5
5	4	1	8	2	6	7	9	3
9	3	8	1	5	7	6	4	2

Puzzle 87

2	1	8	6	5	7	4	3	9
5	3	9	1	8	4	6	7	2
4	6	7	2	3	9	5	1	8
6	9	5	4	1	2	3	8	7
3	7	2	5	9	8	1	4	6
1	8	4	3	7	6	9	2	5
9	2	6	8	4	1	7	5	3
7	4	3	9	2	5	8	6	1
8	5	1	7	6	3	2	9	4

Puzzle 88

2	8	9	7	1	6	3	5	4
6	5	3	8	9	4	2	1	7
7	4	1	2	3	5	6	8	9
8	6	5	4	7	3	1	9	2
3	7	4	1	2	9	5	6	8
1	9	2	5	6	8	7	4	3
5	3	6	9	4	7	8	2	1
4	1	7	6	8	2	9	3	5
9	2	8	3	5	1	4	7	6

Puzzle 89

3	1	2	8	5	9	7	6	4
6	4	9	1	7	3	5	2	8
8	7	5	4	2	6	1	3	9
1	3	8	6	4	7	9	5	2
2	5	6	3	9	8	4	1	7
7	9	4	5	1	2	6	8	3
4	8	7	2	6	5	3	9	1
5	2	1	9	3	4	8	7	6
9	6	3	7	8	1	2	4	5

Puzzle 90

6	8	5	7	2	4	9	1	3
1	7	2	9	3	5	8	4	6
4	9	3	1	6	8	2	7	5
9	2	8	5	4	6	1	3	7
3	4	6	2	7	1	5	9	8
5	1	7	3	8	9	6	2	4
2	5	4	6	9	3	7	8	1
7	3	1	8	5	2	4	6	9
8	6	9	4	1	7	3	5	2

Puzzle 91

5	1	9	4	6	7	2	8	3
2	7	8	9	3	5	1	4	6
4	3	6	8	1	2	7	5	9
6	2	4	7	8	9	5	3	1
1	8	3	5	4	6	9	7	2
7	9	5	3	2	1	4	6	8
3	5	7	1	9	8	6	2	4
8	6	1	2	5	4	3	9	7
9	4	2	6	7	3	8	1	5

Puzzle 92

8	1	4	5	6	7	3	9	2
5	7	9	2	3	8	1	4	6
3	6	2	4	1	9	8	7	5
2	3	6	8	9	1	7	5	4
4	9	8	7	2	5	6	1	3
1	5	7	3	4	6	2	8	9
6	8	3	1	5	4	9	2	7
9	4	1	6	7	2	5	3	8
7	2	5	9	8	3	4	6	1

Puzzle 93

1	3	8	2	4	5	6	7	9
5	6	2	8	7	9	1	3	4
4	9	7	1	6	3	5	2	8
9	5	4	3	1	7	2	8	6
7	1	6	5	2	8	4	9	3
2	8	3	6	9	4	7	1	5
3	4	5	7	8	1	9	6	2
8	2	1	9	5	6	3	4	7
6	7	9	4	3	2	8	5	1

Puzzle 94

4	3	9	2	8	6	5	1	7
6	2	8	7	5	1	3	9	4
7	1	5	4	9	3	2	6	8
2	8	3	6	1	9	4	7	5
1	9	7	5	2	4	8	3	6
5	4	6	8	3	7	1	2	9
8	6	2	1	7	5	9	4	3
3	5	4	9	6	2	7	8	1
9	7	1	3	4	8	6	5	2

Puzzle 95

2	4	7	1	3	5	9	8	6
5	8	1	2	9	6	7	3	4
6	9	3	8	4	7	5	1	2
3	1	4	5	8	2	6	9	7
7	2	8	4	6	9	3	5	1
9	6	5	7	1	3	2	4	8
4	5	9	6	2	8	1	7	3
1	7	6	3	5	4	8	2	9
8	3	2	9	7	1	4	6	5

Puzzle 96

3	4	6	8	7	5	2	1	9
5	2	7	3	1	9	4	6	8
9	8	1	6	2	4	5	7	3
7	5	9	4	3	8	1	2	6
4	1	8	2	9	6	7	3	5
2	6	3	1	5	7	9	8	4
1	7	4	5	6	3	8	9	2
6	9	5	7	8	2	3	4	1
8	3	2	9	4	1	6	5	7

Puzzle 97

8	2	7	1	5	3	4	6	9
3	6	9	4	7	2	1	5	8
4	1	5	9	8	6	3	2	7
9	4	6	5	1	7	8	3	2
7	8	3	2	6	9	5	4	1
1	5	2	8	3	4	9	7	6
5	9	4	6	2	1	7	8	3
2	7	8	3	9	5	6	1	4
6	3	1	7	4	8	2	9	5

Puzzle 98

6	2	8	3	9	4	1	7	5
7	4	5	1	2	6	8	3	9
3	9	1	8	5	7	2	6	4
4	7	2	6	1	9	5	8	3
1	3	6	2	8	5	9	4	7
5	8	9	4	7	3	6	2	1
8	1	7	9	3	2	4	5	6
9	6	3	5	4	8	7	1	2
2	5	4	7	6	1	3	9	8

Puzzle 99

4	3	5	8	7	1	6	2	9
8	9	2	6	4	5	7	1	3
6	1	7	3	2	9	4	5	8
7	8	4	2	9	3	1	6	5
1	5	3	7	6	4	8	9	2
9	2	6	5	1	8	3	4	7
3	4	8	9	5	6	2	7	1
2	6	9	1	3	7	5	8	4
5	7	1	4	8	2	9	3	6

Puzzle 100

6	3	9	2	7	1	4	5	8
5	4	2	3	8	9	1	7	6
1	7	8	4	6	5	2	3	9
3	6	7	1	9	4	5	8	2
9	1	4	5	2	8	7	6	3
2	8	5	7	3	6	9	4	1
4	5	3	6	1	2	8	9	7
7	9	1	8	5	3	6	2	4
8	2	6	9	4	7	3	1	5

www.ingramcontent.com/pod-product-compliance
Lightning Source LLC
Chambersburg PA
CBHW070818220526
45466CB00002B/712